胡同门楼建筑艺术

The Architectural Art of Hutong Gate Buildings

李明德 著
李海川 摄影
Written by Li Mingde
Photographed by Li Haichuan

中国建筑工业出版社
China Architecture & Building Press

本书以胡同中四合院门楼建筑，及其门楼、门头、门枕石、上下马石、影壁等造型艺术为例，图文并茂地展示其高超的建筑艺术和民族文化。在门楼建筑上，装饰着砖雕、石雕、木雕，其造型多样 、画面丰富、工艺高超。虽历经沧桑，至今仍保留下众多的艺术精品，为研究、考证胡同文化提供了生动的依据。

前言 Forword

北京是一座世界闻名的文化古都，有着三千多年的建城史。

在丰富悠久的历史文化中，胡同建筑占有着重要地位。据史料记载，一些古老街巷形成于元代1276年起，经历明、清沿袭至今，大部分建筑及胡同的位置均少有大的变化。

本书仅以胡同中四合院门楼建筑，及其门楼、门头、门枕石、上下马石、影壁等造型艺术为例，通过简要的文字介绍，结合多幅照片，展示其高超的建筑艺术和民族文化。在门楼建筑上，装饰着砖雕、石雕、木雕，其造型多样、画面丰富、工艺高超。虽历经沧桑，至今仍保留下众多的艺术精品，为研究、考证北京胡同文化提供了生动的依据。每当人们穿街过巷，留心观赏，那一座座造型优美的门楼和装饰在上面的砖、石雕刻精品，会得到高度的艺术享受。

北京在前进。城市现代化正加速进展的时刻，人们对胡同的历史、文化、民俗却有着深深的眷恋之情。本书展示着古老胡同中门楼建筑的画面，从历史和文化艺术方面进行分析介绍，并对所附百余幅照片加以文字说明，以供广大读者研究欣赏。从而进一步认识其历史价值，弘扬它辉煌的一页。

With a history of more than 3000 years Beijing is a world—renowned, cultural city.

In its rich and prolonged history and culture construction of alleys or Hu tongs occupies an important position. According to historical records some of its ancient streets and lanes took shape during the Yuan dynasty (beginning from 1276 A. D.) and were carried on to the present through the Ming and Qing dynasties. Most of the buildings and locations of the alleys have not undergone much changes.

Taking the gates of some quadrangles in Beijing's alleys and the decorative arts on the gates, gate lintels, gate piers stepping stones for mounting or dismounting from a horse and screen walls as examples this book attempts to show the superb, architectural art and folk culture of the gates through brief words and many colored pictures. The varied, rich and fine brick, stone and wooden carvings on the buildings of the gates remain a treasure of works of art today despite many vicissitudes. They provide vivid evidences for the study and textual research of the culture of Beijing alleys. Whenever people walk through the alleys and look at the beautifully—constructed gates and the decorative brick and stone carvings on them they feel they are enjoying a high degree of artistic beauty.

Beijing is fast advancing toward modernization. Still people cherish a profound love for its alleys with their history, culture and folk customs. With a historical and artistic analysis and introduction of the gates buildings in the alleys and with over a hundred illustrated pictures this book will enable our readers to fully appreciate and further realise their historical values and to carry forward their brilliant aspects.

目录 Contents

丰富多彩的门楼
Colorful Gate Buildings

在北京的胡同里，四合院的门楼是最引人注目的建筑。那一座座造型各异、大小不等的门楼，展示着高超的营造艺术和京城悠久的传统文化。早在元代扩建大都时，遵循《周礼·考工记》中所规定的“左祖右社、面朝后市”的原则设计。对城市的街巷按统一的标准来兴建，城内分五十坊，坊与坊之间形成平直的大街小巷。据明《永乐大典》所辑《析津志》载：大街宽二十四步、小街宽十二步、胡同宽六步（一步为今天的1.54m）。大的街巷为南北方向，一般胡同则沿着南北大街的东西侧顺序排列。明永乐四年（公元1406年）在元大都基础上又进行了扩展。历经明、清两代乃至民国和解放以后北京大部分胡同的布局建置均无太大的变化。今天，北京胡同约有4000余条。民间流传着顺口溜儿：“有名的胡同三千九，无名的小巷赛牛毛。”一些老街巷，仍保持原位置和名称。如：地安门东大街的南锣鼓巷地区就是很好的实例。元代这里称昭回坊与靖恭坊。那南北的主街、东西对称的十余条胡同，形成了蜈蚣状，俗称为蜈蚣巷。胡同内不同类型的四合院门楼很齐整，一些老宅院门前还保存着“上、下马石”，大门两侧还配有造型精美的“抱鼓石”，以及影壁、老槐树等景观，其中部分胡同还是原来的名字。这是胡同里古建筑与民俗文化的真实写照。在清代鼓词说唱中，对宅第的门楼如此描述：“青水脊的门楼安着吻兽，两边有两溜拴马的桩。上马石下马石分为左右，铁丝儿的灯笼挂在当央。”北京民间还流传着：“宅子老不老，要看槐树大和小”的谚语。其意是通过对四合院门楼前种植槐树的树龄来验证古老门楼建筑的年代，这还是有说服力的。一般大四合院门楼前左右各种槐树一株，其种植与建门楼同时进行的。四合院不仅是我国古建筑中的典型，门楼的样式、规格因宅第主人的身份而区分。采用坐北朝南的位置，在敞亮的院子里，北面为正房，东西为厢房，南屋称为倒座，门楼大部分均建在院东南角。按古代风水学的要求，“乾宅巽门”为最佳位置。“巽”则指东南，故门楼建在四合院的东南方，以示吉利。北京古老胡同里现存的四合院门楼多为清代所建，至今有一二百年的历史，也有极少数是明代的建筑。这些年代悠久、造型优美、不同规格的门楼，包括其门头、戗檐、门簪、门墩儿……的装饰艺术，有着高度的研究、欣赏价值。为今天我们了解北京胡同建筑历史及文化、民俗提供了活化石，是多么难能可贵啊！

在封建社会中，门楼显示着主人的品位等级。建筑规格及造型装饰是有所区分的。清代的王府门楼很壮观，一般是坐北朝南。有五间三启门和三间一启门。如：朝内大街的孚王府，系清康熙皇帝第十三子的府第，正门五间，门前左右各立大石狮一个。后海的恭王府正门三间，门前石狮一对。这样保存完整的府第今天已为数不多了。官宦及富户的门楼多为广亮大门，位置在宅院的东南角。其造型高大，门头及戗檐装饰精美的砖雕，台基也高大。门外广亮宽敞。有些府第的广亮大门门楼，左右两侧砌八字粉墙（俗称：撇山影壁），上为筒瓦，墙壁磨砖对缝。门楼与粉墙之间留有宽阔地面，设置上下马石一对，其倒座墙上配有四个拴马桩，更显示其门户整体的端庄和严肃。这样的门楼，今天在老胡同中还能见到。但大多数广亮大门前是不砌八字粉墙的，更不能设置上、下马石。一般中小型四合院多为如意门。在北京数千条胡同中，如意门楼的数量很多。其建筑特点是大门设在外檐柱间，门框两侧砌砖墙。门楣上有较精细的砖雕图案，门楣与两侧砖墙交角处砌如意状的砖饰，表示“吉祥如意”的意思，俗称：如意门。街巷中的如意门楼多为清代建筑，也有部分是民国期间所建的。大如意门，门楼占一整间。中型如意门，门楼占多半间。小型如意门，门楼占半间。在为数众多的如意门中，门楼建筑造型上是多彩多姿的。有的砖料上乘，砌工考究，门头栏板及戗檐的砖雕图案内容丰富，门簪上的木雕工艺精巧，有文字或花卉图案。有的门上还钉有对称的铁皮图案。这些建筑上的装饰都反映出我国古代劳动人民的智慧，其中精品有着高度研究价值。还有为数不少的墙垣式门，其规格较小。门楼与院墙相连，一般较小的四合院或三合院多采取这样的建筑形式。另外，在胡同里老北京平民百姓所居住为数最多的是随墙门，其造型很简易，然而又最为普遍。有的门墙用碎砖头和泥灰砌成，北京人称这类碎砖墙为“核桃酥”，比喻其建筑质量之差。

The most attractive buildings in Beijing's alleys are the gates of quadrangles. Of various shapes and sizes these gates display China's superb architectural art and Beijing's long-standing traditional culture.Back in the Yuan dynasty when Dadu (present-day Beijing) was being expanded the designing principle followed was "to build the ancestral temple on the left side and site of sacrifices to the God of land on the right side, with the court in front and market at the rear" laid down in " Zhou Li" or "the Book of Rites of the Zhou Dynasty" . The city's streets and lanes were constructed according to a unified standard. There were 50 Fang or portions of land, with straight streets and lanes between them. According to the Yongle Canon compiled during the Ming dynasty, an avenue was 24 Bu(one Bu is equivalent to 1.54 metres) wide, a street was 12 Bu wide and an alley 6 Bu wide. Big streets and lanes were south-north oriented, ordinary alleys ran from east to west on the flanks of the streets. Further expansion was carried out on Dadu in the fourth year of Yongle reign of the Ming dynasty (1406 A.D.). From that time on down through the Ming, Qing dynasties, the Republic of China (1912~1949) and post liberation days the layout of the great part of Beijing has not been changed much. Today Beijing has over 4000 alleys. A popular doggerel goes, "Well-known alleys number 3900 and unknown alleys are countless just like cow hair ." Some old streets and lanes retain their orginal locations and names. For instance, the Nanluogu Xiang district of the Di'anmen Dongdajie (street). During the Yuan dynasty it was called Zhaohui Fang and Jinggong Fang. The north-south oriented avenue and over a dozen symmetrical west-east oriented alleys present the shape of a centipede, hence popularly known as Wugong Xiang or the centipede lane. The lane boasts rows of neat quadrangles with their variegated shapes of gates. Some time-honored dwellings still preserve stepping stones for mounting or dismounting from a horse in front of their gates, exquisitely skulpted Drum Stones on both sides of their gates as well as screen walls, locust trees and other scenic sights. Part of the alleys still use their original names. These give a true portrayal of ancient buildings and folk customs of Beijing's alleys. In story-telling and ballad-singing of the there was such description of the gates of Beijing quadrangles: "Skulpted animals sit on the bluish ridges over the gates ,on two sides are posts for tethering horses, on the right amd left are stepping stones for mounting or dismounting from a horse and lanterns are hung in the centre." A folk saying runs," To determine the age of a residence, just look at the size of the locust tree." It means that you can determine the age of the residence through the age of the locust trees planted in front of its gate. Generally a locust tree is planted on either side of the gate. They are planted at the same time when the residence is constructed. Quadrangles are a typical example of ancient Chinese buildings. More than that the styles and standards of their gates vary in keeping with the social status of their owners. They were generally built facing south with spacious and well-lighted courtyards. The principal room faced south, wing rooms were on the west and east sides and the room facing north is called Dao Zuo Fang or reversely-set room. Most of the gates were opened on the south-east corners of the courtyards. According to the theory of geomancy, it is the best and most propitous location. Existing quadrangles of Beijing's old alleys and their gates were mostly built in the Qing dynasty with a history of one to two hundred years. A small number of them were built in the Ming dynasty. These time-honored, beautifully-shaped and varied styled gates together with their decorations such as lintels, Qiang Yan or eaves, clasps, piers and others are highly valuable for both research and artistic appreciation. They are living fossils providing us with clues to study the history, culture and people's customs of buildings in Beijing alleys.

In the feudal society the gates demonstrated the ranks and social classes of their masters. They were differenciated in constructional standard and sculptural decoration. The gates of residences of Princes of the Qing dynasty were magnificent. They usually face south with 5-bay across or 3-bay across. For instance, Fu Wang Men or the residence of Prince Fu, the 13th son of Emperor Kangxi of the Qing dynasty at Chaonei Dajie. It has a 5-bay principal room having a pair of big stone lions on the two flanks of its gate. Gong Wang Fu or the residence of Prince Gong at Hou Hai or Rear Sea has a 3-bay principal room with a pair of stone lions in front of its gate. Such well-preserved residences are not many. The dwellings of official and wealthy families mostly have Guang Liang gates located on the southeast corners of the houses. These gates are big and tall, their lintels and eaves are decorated with exquisite brick carvings. They also have high steps and spacious areas outside. Some of these gates are flanked with white-washed walls shaped like the Chinese character "八", (popularly known as Pie Shan screen walls). They are topped with tiles and linked up at the back. Broad spaces are left between the gates and white-washed walls where a pair of stepping stones for mounting and dismounting from a horse are installed. There are also 4 posts for tethering horses. All this adds dignity and solemnity to the gates. Such gates can still be found in old-time alleys today. But most of the Guang Liang Gates do not have white-washed walls, let alone stepping stones for getting on or down from a horse. Ordinarily medium-sized and small quadrangles have Ru Yi Gates. Their number is the greatest in several thousands of Beijing alleys. They are opened between outer columns and have brick wall on both sides. On their lintels, there are fine patterns of brick carvings. The juncture of lintels and brick walls decorative Ru Yi-shaped bricks implying auspiciousness. Thus people call them Ru Yi Gates. Ru Yi Gates of Beijing' streets and lanes were largely built in the Qing dynasty, part of them were constructed during the Republic of China (1912~1949). Big Ru Yi Gates are one-bay across, medimm-sized ones occupy more than half bays and small Ru Yi Gates take up half a bay. Buildings of the numerous Ru Yi Gates are rich and colorful in architectural design and style. Some have meticulously-laid, fine-quality bricks, their panels and eaves bear a great variety of carved brick patterns and their gate clasps are decorated with fine wood carvings with words or floral patterns. Some Ru Yi Gates are set with symmetrical patterns of iron sheets. All these constructional decorations show the wisdom of the working people in ancient China. Exquisite works among them are of great value for research work. There are also a good number of wall-typed gates, comparatively smaller. Linked with walls of courtyards these gates are generally adopted by smaller quadrangles or San He Yuan, threeroom compounds. Besides there are Sui Qiang Gates adopted by the majority of old Beijing residents. They are simple and widespread. Some Sui Qiang Gates are built of broken pieces of bricks and marl. Beijingers call these gates "Walnut shortbreads", a metaphor insinuating their poor quality.

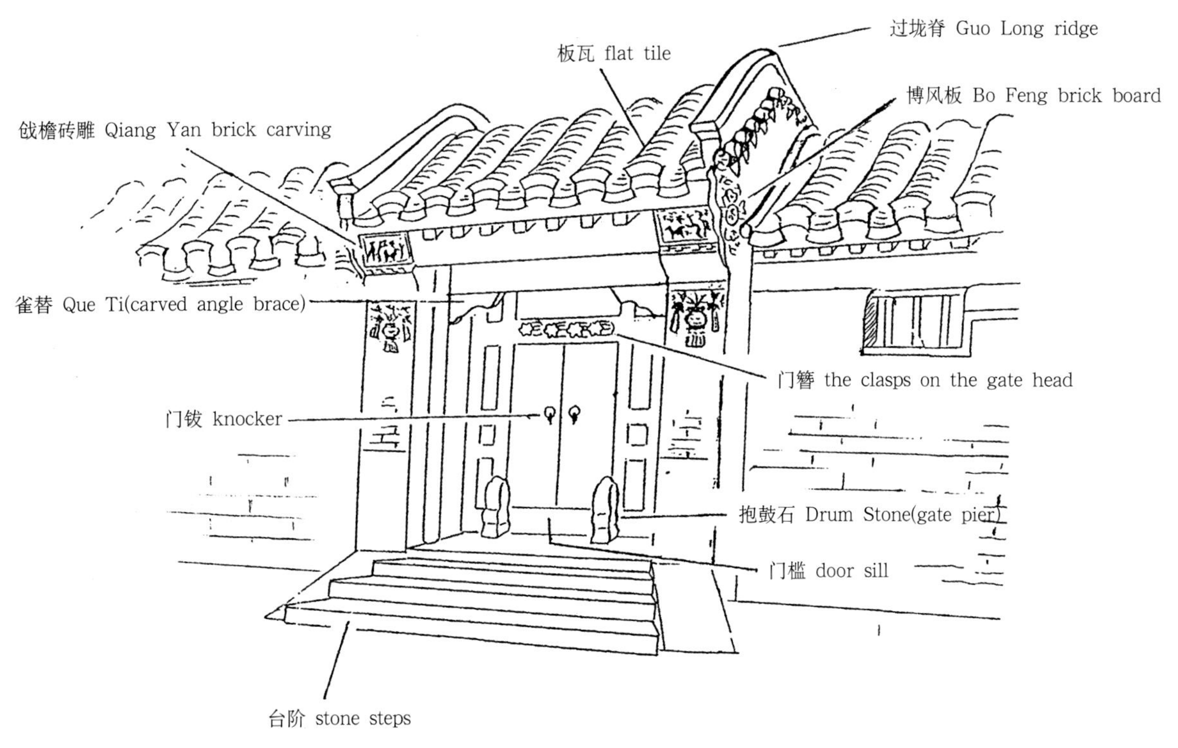

广亮大门门楼建筑结构图

The construction of the Guang Liang Gate building

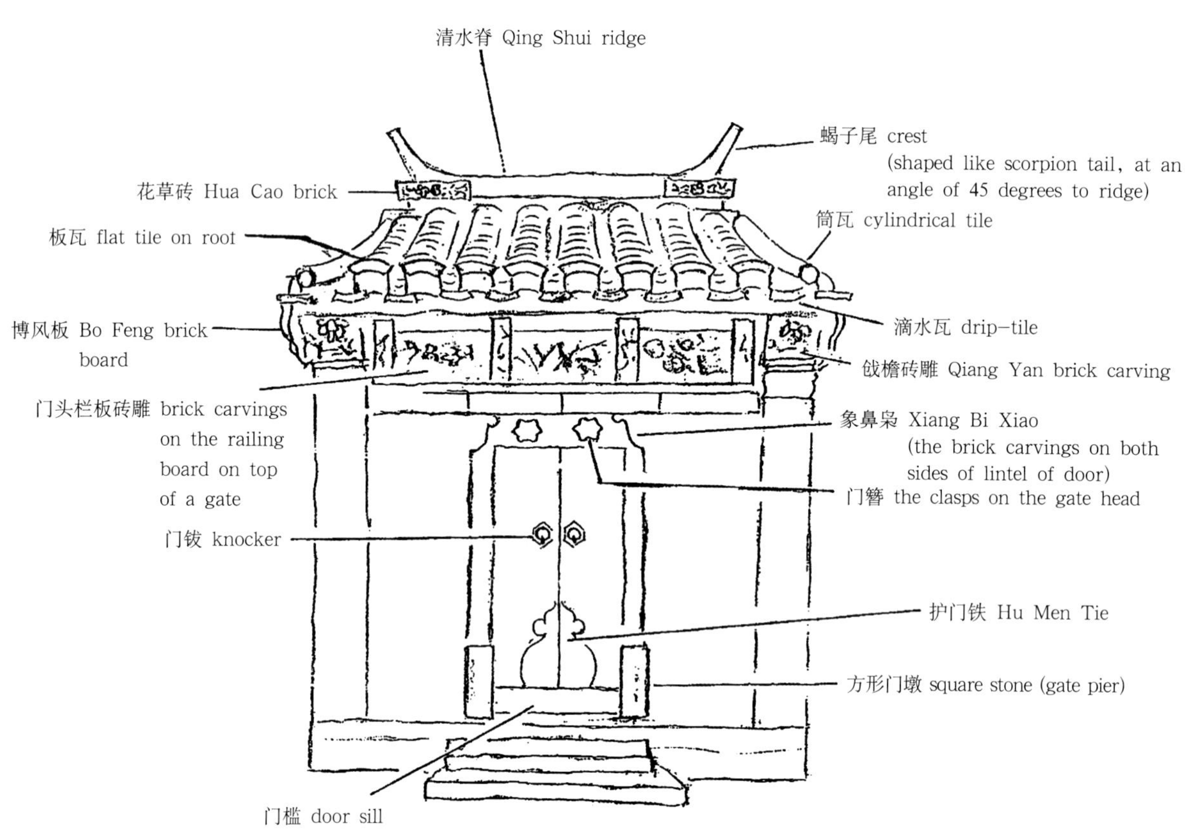

如意门门楼建筑结构图

The construction of the Ru Yi Gate building

王府的门楼多为五间三启门或三间一启门

The gates of residences of princes are mostly 5-bay across with three pairs of doors or 3-bay across with one pair of doors

现存完好的四合院广亮大门门楼，造型精美，砖雕工整，显示出宅主人的品级地位。

A well-preserved gate of a large quadrangle. Exquisitely-designed brick carvings show the social position of the house owner.

这是蛮子门，大门的门扉安装在外檐柱间就没有大门洞了，是广亮大门的另一种形式。
This is a Man Zi Gate, another type of Guangliang Gate. With its panels set between outer columns it has no gateway.

为数众多的如意门门楼
Numerous Ru Yi Gate buildings

建于不同年代的老门楼

Gate buildings built in the different periods

造型多样的老门楼

Various old gate buildings

砖雕精美的老门楼

The old gate building with beautiful brick carvings

建于不同年代的西洋式门楼
Western-styled Gates Buildings built in the different periods

配置拴马桩的洋式门楼
Western-styled Gates Buildings equipped with the post to tie horse

有方形门墩的洋式门楼
A Western-styled Gates Buildings equipped with a pair of square stones

精美的门楼砖雕
Exquisite Brick Carvings on the Gates

北京胡同四合院门楼上的砖雕、石雕装饰，有着高度的观赏价值，独具艺术光彩。门头上的栏板及两旁戗檐上的砖雕图案不仅内容丰富，其雕刻技法高超。虽经过百年以上的风雨侵蚀，这些精美佳品至今仍保存完好，与整体门楼建筑结合得恰到好处，形成极好的装饰效果。追溯北京砖雕艺术的历史，以明清时代最为兴盛，那时府第、会馆、民居的四合院门楼上装饰砖雕、石雕很普遍，用于门头、门墩儿、屋脊等处。 目前，现存一些老街巷门楼上保存的砖雕，多为清代工匠所雕，也有部分是民国期间的砖、石雕品。工匠们采用薄肉雕、浮雕、透雕和线刻等多项高难的技法，根据门楼宅第的不同，在门楼、门头、戗檐、门墩儿上进行图案设计，按照图样的尺寸去烧制澄浆泥砖，在砖上雕刻。其中以浮雕最具特色，有些画面的精彩部分，则单独制做，后再镶在砖面上，如：花朵、兽头……。还有部分运用透雕技术，表现出的图案效果更佳。用拼接的方式，如两拼、四拼、六拼、八拼等完成门头的整个砖雕装饰。画面上呈现的人物、花卉图案的完成，要靠工匠的刀工技巧，一般雕完后不再另行打磨，几块砖雕拼砌在一起，就形成了巨幅的佳品了。今天，北京会这门砖雕技术的老艺人已寥若晨星。

门头（或称：门额）、戗檐上的砖雕精品，有着浓郁的地方特色，是胡同文化史与民俗的写照。细心观赏，能分辨出画面的丰富内涵。砖雕的内容大致可以分为如下几类。

1.吉祥富贵的图案。展示人民对美好幸福生活的向往。以蝙蝠、鹿、喜鹊、仙鹤、麒麟、盘长、如意、磬……分别组成图案，各有其讲法和寓意。有的图案还与其谐音联系在一起，是很有讲究的。在栏板浮雕蝙蝠（谐“福”音）、鹿（谐“禄”音）、古代乐器磬（“磬”即是喜庆的谐音）。“喜庆吉祥”的图案中，画面上的蝙蝠口叼着磬，表达了主人求喜庆，盼福到的心愿。“福寿绵长”图案中，蝙蝠口叼盘长（佛八宝之一）其图案中绸带不封口，表示佛法无边，其寓意是福寿绵长。“鹤鹿同春”图案，是表示延年益寿。

2.展示花卉的图案。如：梅花、兰草、翠竹、菊花，这是我国的传统名花，称为“园林四君子”。自古以来就得到文人墨客的吟咏。宋代文学家王安石就留有赞梅花的诗：“墙角数枝梅，凌寒独自开。遥知不是雪，为有暗香来。”松、竹、梅组成图案砖雕，称为“岁寒三友”。还有雕刻石榴、葫芦、葡萄图案的，其寓意是多子多孙，人丁兴旺。

3.“博古”图案。画面内容丰富。将古代器物香炉、玉佩、笔筒、砚台、花瓶……巧妙安排在一组“多宝阁”画面中，表示我国文人的生活。文房四宝与花瓶在一起，寓意四季平安。

4.民间传说与神话故事。经过考查，现在这类人物故事题材的砖雕作品为数极少了。我国古老的“八仙过海”故事，将韩湘子、张果老、李铁拐、曹国舅、蓝采和、汉钟离、吕洞宾、何仙姑的传说，活生生展现在画面上，是北京砖雕艺术中的佳品。表现形式上，又分明八仙和暗八仙两种，韩湘子吹笛、曹国舅持玉板高歌、李铁拐祝寿、何仙姑采莲、张果老骑驴……这类人物图案为明八仙。在图案上仅雕刻八仙常佩带的器物，如：韩湘子吹的笛子、曹国舅手中的玉板、蓝采和的花篮、汉钟离的扇子、何仙姑手持的荷花、李铁拐的葫芦、吕洞宾的宝剑、张果老的渔鼓配之以祥云、绸带、荷叶组成图案的，均为暗八仙。京城有一处门头砖雕，名为“竹林七贤”。上有象征吉祥的佛八宝图案，即宝伞、胜利幢、宝瓶、双鱼、莲花、海螺、吉祥结、金轮，更是珍品，展现魏晋年间，阮籍、山涛、刘伶、向秀、王戎等七位文人名士经常游于竹林饮酒赋诗，抚琴吟唱的情景，画面生动感人，是目前胡同砖雕仅存的精品。其他如“麒麟送子”“马上封侯”等图案，内容可谓丰富多彩。

5.亭台楼阁图案。

6.福字、寿字、云纹等图案。门楼左右上方的戗檐，砖雕图案多为花卉、动物或人物故事。如：雕刻兽中之王狮子的图案，象征着宅主人的武官身份。雄狮戏球的画面更为生动活泼，这里还有太狮、少狮之分。还有浮雕富贵牡丹或牡丹花篮、松鹤同春等花卉图案的。戗檐上的人物砖雕更为罕见。这些图案构图精美、生动，刀工玲珑剔透。

屋脊与房瓦 广亮大门与如意门门楼上的房瓦既坚固又美观，有些为板瓦组成的合瓦，有些为筒瓦。一些房瓦还雕刻有“吉祥”、“喜”、“寿”等文字以及盘长、蝙蝠、荷花等图案。尤其是一些老宅门历经百年风雨，至今屋脊和房瓦保存完好。门楼屋脊也是很有讲究的，分清水脊和过垅脊两种。清水脊在屋脊两端一左一右对称翘起，斜伸对天，脊长40～60cm，名为“朝天笏”，俗称“蝎子尾”。下座配花草砖，砖雕图案多为松竹梅、牡丹等。中小型门楼清水脊所配的花草砖工艺相对简单。过垅脊门楼为数也不少，即在门楼顶部两端砌脊，两端没有翘起，从侧面看顶部呈“人”字形。

Brick and stone carvings on the gates of quadrangles in Beijing alleys are highly valuable for ornamental purpose and possess unique, artistic splendor. The carved brick patterns on the lintels and on Qiang Yan or eaves on both sides are not only rich in content but also superb in carving skill. Though weatherbeaten for over a hundred years the cream of them are still preserved. Well matched with the whole gate buildings they achieve a good decorative effect and are loved by the people. The history of the art of Beijing's brick carving can be traced back to the Ming and Qing dynasties when it was flourishing. At that time it was popular to mount brick and stone carvings on the lintels, piers and roof ridges of the gates of quadrangles of civilian dwellings, guild houses and officials' residences. Most of the extant brick carvings were made by artisans of the Qing dynasty, part of the brick and stone carvings were done during the Republic of China (1912~1949). Applying various highly difficult carving skills such as thin carving, bass relief carving, open work carving and linear carving, the artisans first drew designs on the arch gateways, lintels, Qiang Yan or eaves and gate piers and baked bricks according to the designed sizes and then made carvings on them. The most outstanding is bass relief carving. Attractive parts like flowers, animal heads and others were made separately before they were inlaid on the bricks. Open work carving technique could achieve the best artistic effect. The whole carved brick decoration of the gate heads was accomplished by joining together 2,4,6,8 and more pieces. It depended on the craftsmanship of artisans to complete the patterns of figurines, flowers and plants. Once carved, usually no further polishing was needed. The joining together of several carved bricks formed a great master-piece. Today only a small number of masters skilled in such brick carving can be found in Beijing.

With a strong local flavor exquisite brick carvings on the gate lintels (or gate heads) and Qiang Yan or eaves are a portrayal of the history of Hutong (or alley) culture and folk customs. At a close look one can understand the rich contents of their designs. They can be divided into the following categories.

1. Designs of auspiciousness, wealth and nobility express people's yearning for a happy life. Patterns composed of bats, deer, magpies, white cranes, Chinese unicorns, Pan Chang Ru

Yi (an S-shaped ornamental object symbolizing good luck), chime stones, etc. Each has its own meaning and implications. Some patterns have something to do with homophony. In bass relief carving on railing boards, bat (Bian Fu) is homonymous with "福" (Fu) or good fortune, deer pronounced Lu in Chinese with "禄" (Lu) or official rank and chime stones pronounced Qing in chinese with "庆" Qing or jubilation. In the pattern of "jubilation and auspiciousness" the bat (s) holds a chime stone in its mouth, expressing the house owner wish for attaining happiness and good fortune. In the design "infinite happiness and a long life" the bat holds in its mouth a Pan Chang, one of the 8 Buddhist treasures ("Chang" in Chinese means long).For the silk ribbon in the picture not being closed implys that the powers of Buddha are boundless and that the owner can enjoy infinite happiness and longevity. The pattern of "crane and deer" expresses the good wish for a prolonged life.

2. Designs of flowers such as plum blossoms, orchids, emerald bamboos and chrysanthemums. These are all China's famous traditional flowers known as "the four gentlemen of the garden" They have been praised in poems by letters men since ancient times. Writer Wang Anshi of the Song dynasty composed a poem eulogizing plum blossoms. The poem goes like this:Several plum blossoms at a wall corner, blooming lonely in defiance of cold; Knowing from afar they are not snow for their fragrance is wafted in the air. The pattern of pine, bamboo and plum blossom on carved bricks is called "the three companions of cold winter". Besides there are carved pomegranates, gourdsand grapes implying bearing many children to make the family tree thrive.

3. Designs of ancient relics. Ancient articles such as incense burner, jade plate, brush pot , inkslab, vase······are ingeniously put together to form a picture of "curio shelf", depicting the life of Chinese scholars. To place the four treasures of the study namely writing brush, ink stick, ink slab and paper with the vase expresses the meaning that things would run smoothly in all four seasons.

4. Folk tales and mythological stories. After investigation only a small number of carved bricks on such subject matters exist today. The best of Beijing's carved brick art are the fairy tale about eight immortals crossing the sea. The 8 immortals are Han Xiangzi, Zhang Guolao, Li Tieguai, Cao Guojiu, Lan Caihe, Han Zhongli, Lü Dongbin and He Xiangu. There are two ways of presentation, open and concealed. The open way is to portray Han Xiangzi playing flute, Cao Guojiu singing with jade clappers, Li Tieguai attending a birthday celebration, He Xiangu picking lotuses, Zhang Guolao riding on a donkey so on and so forth. The concealed way is to carve only the articles the 8 immortals usually carry such as the flute of Han Xiangzi, the jade chappers of Cao Guojiu, the floral basket of Lan Caihe, the fan of Han Zhangli, the lotus of He Xiangu, the gourd of Li Tiegua, the sword of Lü Dongbin and the Yu Gu, a percussion musical instrument made of bamboo, decorated with clouds, silk ribbons and lotuses. Brick carvings on the gate heads entitled,"7 outstanding scholars of bamboo grove" are also gems of art that have been left. It have patterns of the eight Buddhist treasures symbolized auspiciousness: Umbrella, Sheng Li Chong (streamer used in ancient China), Vase, Double Fish, lotus flower, Ji xiang jie (the Chinese Knot), Jin Lun (the pattern shaped like wheel). They describe scenes about 7 men of letters,including Yuan Ji, Shan Tao, Liu Ling, Xiang Xiu and Wang Rong during the period of Wei and Jin who composed and recited poems, played lutes and drank wine while roaming in bamboo groves. The portrayals are vivid and moving. Other designs about folk tales and fairy stories include "The unicorn brings a baby" and "returning the granted title and seal".

5. Designs of pavilions and chambers.

6. Designs of Chinese characters such as for fu meaning

good fortune, shou meaning longevity and yun meaning clouds. Carved brick patterns on Qiang Yan over the right and left sides of the gates are largely flowers, animals or characters in novels. For example, the pattern of lion and the king of animals indicates the military rank of the house owner. The picture of lions playing with a ball is lively and vivid. In the pictures there are young lions as well as grown ones. Floral patterns carved in bass relief include peony or peony basket symbolising wealth and rank and "pines and cranes" signifying longevity. Brick carvings of characters on Qiang Yan are a rare sight.

Ridge and tile of the house. Carved on the gate tower of Guangliang Gate and Ru Yi Gate, the house tiles are both firm and beauty, some of which are plain tile composed of pan tile and some of which is composed of pantile. Some house tiles are also carved with character such as "吉祥"(means auspiciousness), "寿"(means longevity) and patterns such as Pan Chang(a kind of Chinese knot symbolizes auspiciousness and good fortune), bat and lotus. Having experienced the erosion of a hundred year of rain and wind, the ridge and house tile of some gate of the old-style still preserved in good condition. The ridge of the gate tower, including two shapes namely Qing Shui Ridge and Guo Long Ridge, are also very dainty. The Qing Shui Ridge, the length of which is 40 to 60 centimeter, are symmetrically located on the left and right end of the ridge of a house, tilting upward to the sky. It is named as"Chao Tian Wu", also commonly named as the tail of scorpion. The QinShui Ridge is decorated with Hua Cao Brick, most the carved pattern of which are pine, bamboo and peony. The Huo Cao Brick craft matched with Qinshui Ridge on the middle and minitype gate tower is relatively simple. There are no little amount of gate tower decorated with Guo Long Ridge, i.e. the ridges are laid on the two end of the top of the gate tower without upward tilting. Seen from the side, the top looks like a font style of "人".

由博古图、三羊开泰、鹤鹿同春图案组成的整体砖雕门楹。局部分3组拍摄。

This is an overall brick carving composed of three parts. The patterns are three goats, the curio shelf, pine and crane.

这是门楹砖雕“博古图”。下至挂落板配有人物和蝙蝠、花卉图案浮雕，局部按3部分拍摄。

The Curio Shelf Picture of brick carving on the upper side of gate (The design portrays Chinese ancient relics, known as “the Curio Shelf”). The design of brick carving under the Curio Shelf Picture is composed of figures, bats, flowers.

独特的拱门砖雕与局部
Unique Brick Carvings on the upper side
and on either side of the arch gate

老胡同里不同年代、造型多样的门楹砖雕图案。
Various patterns of brick carvings on the gate head in the old Hutongs.

门头栏板砖雕松、竹、梅、兰图案。

A set of brick carvings on the railing board on top of a gate with patterns of pine, bamboo, plum blossoms and orchids.

一组年代久远的人物砖雕精品。这在京城胡同中是极为罕见的。

The brick carvings of the figures patterns has a very long history. It's rarely seen in the Beijing's alleys nowadays.

如意门门头为数很少的石板雕，造型古朴、大方，独具特色。

The Railing Stone Board on top of Ru Yi Gate is terse, unique and rare.

浮雕多种花卉图案的门楹
The relief brick of the various flowers patterns on the gate head

砖雕精美的宝瓶与花卉的门楹
The brick carving of the vases and flowers patterns on the gate head

透雕“八骏图”及花卉的门头
The pattern of eight horses and flowers carved vividly

现保存完好的大型如意门砖雕全貌。其图案表现古代器物，俗称“博古图”。

This is an overall picture of brick carvings on a large Ru Yi Gate found in an old alley The design portrays ancient relics, popularly known as “the Curio Shelf”.

不同图案的门楹砖雕
The brick carvings of different patterns

清代官宦之家门楼戗檐砖雕。上边浮雕“鹤鹿同春”，下面是“富贵牡丹”花篮图案。

These are brick carvings on Qiang Yan (eave) of the gates of official families of the Qing dynasty. The upper part is the bass relief pattern “Crane and deer” while the lower part is the pattern of a floral basket entitled, “Peony, the symbol of wealth and nobility”.

戗檐与博风板的砖雕图案

The patterns of brick carvings on the Qiang Yan and Bo feng Ban

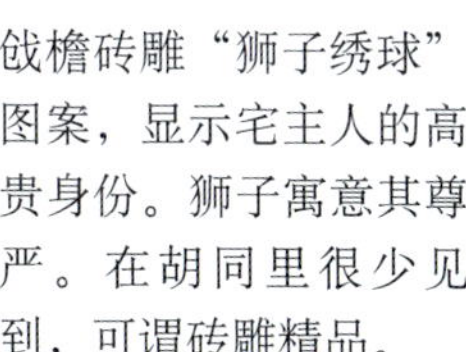

戗檐砖雕“狮子绣球”图案，显示宅主人的高贵身份。狮子寓意其尊严。在胡同里很少见到，可谓砖雕精品。

The pattern, “Lions play with a ball”, indicates the high rank of the house owner. Lion symbolizes dignity. It’s rarely seen in the alleys nowadays so it is a gem of brick carvings.

砖雕麒麟图案，是吉祥与智慧的象征。
The pattern of “Unicorn” symbolize the wisdom and auspiciousness.

透雕“博古图”，展示宅主人的文化品味。
The brick carving of the curio shows the house owners’ taste for culture.

广亮大门门楼上一左一右的戗檐以山羊、猴子、花草组成的画面，生动活泼，透雕精美。

Brick carving on the Qiang Yan on the both sides of gate head composed of goats, monkeys, glass.

“马上封侯”图案
The pattern named “Ma Shang Feng Hou”(It’s the Chinese traditional subject to symbolize granted title composed of horse monkeys)

“犀牛望月”图案
The pattern of a bull viewing the moon

透雕“青竹翠鸟”图案
The open work carving of the bamboos and bird

古朴的“松鹿图”
The pattern of primitive simplicity of pine and deer

如意门门楼戗檐，砖雕篆字“延年益寿”，可谓胡同文化中的艺术珍品。

A relief brick carving on Qiang Yan composed of four Chinese characters “延年益寿”and bats, clouds and others. The design means longevity.

葫芦、葡萄、石榴透雕，其寓意：多子多孙、人丁兴旺。

A open work carving with the patterns of gourds, grapes and megranates, implying having greater numbers of descendants or an ever growing family.

由荷花、翠竹、牡丹、菊花组成的戗檐砖雕。

Brick carving on the Qiang Yan composed of lotus flowers, emerald bamboos, peonies and chrysanthemums.

如意门门楼象鼻枭部位的精美砖雕
The brick carving on the Xiang Bi Xiao on both sides of the lintel of a door of Ru yi Gate

清水脊与花草砖

Qing Shui Ridge and Huacao Brick

图案丰富的屋瓦
Colourful patterns of tile

门枕石的雕刻艺术

The Carving Art of the Gates'Pillow-Stones

在欣赏门头砖雕的同时，还要注意门槛两旁的石墩儿，建筑学上称为“门枕石”。这些石墩儿，俗称“门墩儿”。北京的民谣中有：“小小子儿坐门墩儿，哭着喊着要媳妇儿，要媳妇干啥？点灯说话儿，吹灯做伴儿，明天早起给我梳小辫儿”的顺口溜儿。这是居住在四合院的儿童，在大门前玩耍的写照，也体现了人们的孩童时代就与门墩儿联系在一起了。胡同现存的门墩儿，石雕造型上分两种。年代久远的为鼓形，称抱鼓石。近百年所雕多为长方形。其中造型设计绝大部分上面雕小狮子，下面的正、侧面雕有多种浮雕图案，其画面丰富多彩，真可谓是颗颗闪光的石雕精品。关于“门墩儿”上的卧式小狮子，是有讲究的。俗话说：“龙生九子不成龙，各有所长”。我国古代关于龙的传说很多，其中说龙生九种，种种不同。这九种中，老大叫蒲牢，好鸣，做了钟上的钮鼻；老二叫囚牛，喜音乐，做了胡琴上的刻兽；老三叫睚眦，好杀，做了刀剑上的吞口；老四叫嘲风，喜欢冒险，做了殿阁上的走兽；老五叫狻猊，好坐，作了佛座骑象；老六叫霸下，能负重，做了石碑下的托座，就是人们俗称的龟；老七叫狴犴，好打官司，做了牢狱大门上的镇压之物；老八叫屃赑，好文，做了石碑两旁蜿蜒而行的饰纹；老九叫椒图，性敏锐好闭，故用于为主人守门，履行“随手闭门”的职责。以上龙生九子，各有所长（参阅《乾隆皇帝传奇》）。椒图是龙的九子之一，在清代《人海记》中“龙生九子”一节也有论证：“椒图，它生性敏锐好闭，故专为主人守门。”门枕石上所雕的小石狮，是龙的九子之一。它千姿百态、刻工精美，受到人们的喜爱。相附在鼓形及方形石墩儿正面及两侧面上的浮雕图案更是花样繁多，匠心独具了。门墩儿在大小规格尺寸上是等级森严的。①王府大门的抱鼓石高为0.80m、宽为0.45m、厚为0.30m；②广亮大门的抱鼓石高为0.75m、宽为0.45m、厚为0.30m；③如意门的抱鼓石高为0.60m、宽为0.29m、厚为0.19m；④平民百姓家的门墩儿高0.59m、宽0.25m、厚为0.19m。王府与百姓之家差异就很大。正、侧面上的雕刻图案，刻工的精细程度上也有所不同。大部分浮雕画面是花卉，如：梅花、菊花、兰草、翠竹，这是我国的传统名花。还有石榴、蝙蝠、玉笛、葡萄、如意等图案。西城一座王府大门的抱鼓石正面及两侧面的浮雕为古代“乐舞人”，图案造型优美，舞蹈姿态生动，给人们以美的享受。在一些老胡同中，还有“福在眼前（钱与前为谐音）”这类的门墩，正面精雕蝙蝠叼古钱的图案，其寓意“福在眼前，财源滚滚”之意，表示了宅主人的意愿。以上叙述的这几种图案，今天在胡同中细心观赏，还是能见到的。我们在胡同中考查时，发现门楼的门墩儿上保持着完整一对石狮子的已为数寥寥。大部分石狮子，在文化大革命破“四旧”中遭了难，有的四合院门前石墩儿仅存一个完整小狮子，或一对狮子均头尾不全，伤痕累累。今天一些老胡同里还保存着极少部分成对完整无损的小狮门墩，可为珍品。

While enjoying the beauty of brick carvings on the gates' heads one should also pqyattention to the gates' piers on both sides of the threshholds. In architecture they are called the gates' pillows. They are popularly known as grtes piers. A Beijing doggerel goes:“ A little boy sits on the gate pier, he cries for getting a wife; getting a wife for what, to have someone to chat with by light and to have a companion after the light is turned off. The next morning she will braid the piglet for me.” This is a good portrayal of children playing in front of the gates of quadrangles and their early relations with gates' piers. As far as stone carving is concerned existing gates' piers of Beijing alleys can be divided into two types. Ancient ones shaped like drums are called Drum Stones while those carved in the recent hundred years are rectagular. The majoriity of carved ptterns on the top of them are small lions. On the front,right and left sides below, there are carved a great variety of patterns in bass relief. Each and every one of them is a shining example of superb stone carving. There are many stories about the small reclining lions on top of the gates' piers. A folk saying goes: the dragon has 9 sons. Though each has a special skill they all haven't become dragons. Ancient China has many tales about the dragon. One of them goes: the dragon has given birth to 9 different species of animals. The eldest is called Pulao who likes making sounds. So he becomes a handle on the bell; the second son is named Qiuniu, he is fond of music and is carved on Huqin, a Chinese musical instrument; the third son's name is Yazi who likes killing and is thus used to decorate swords and knives; the fourth son is named Chaofeng who loves adventures and so he is used to sit high on the roof ridges of pavilions and halls; the fifth son Suanni who likes sitting becomes an animal on Buddha's pedestral; the sixth son, Baxia, popularly being known as tortoise, is capable of bearing heavy things. He serves as the base on which stone tablets or monu-ments stand. The seyenth son, Bi'an is indulged in making lawsuits. Thus he becomes an object of suppression over the gate of prison; the eighth son called Xibi loves literature and is used as undulating clouds to embellish both sides of stone tablets; the ninth son, Jiaotu, is quick to respond and likes closing, so he seryes as a doorkeeper performing the duty of " closing the door after entering" The 9 sons of the dragon each has his own peculiar capability. (According to references from " Legends about Emperor Qianlong") Jiaotu is the ninth son of the dragon. There are records about him in a chapter on the 9 sons of the dragon from a book of the Qing dynasty enti-tled, "Ren Hai Ji" or "On the Sea of Man" Jiaotu is agile and fond of closing, thus serving as a doorkeeper for his master" The carved little stone lions on the gates' pillow stones are one of the dragon's 9 sons. Meticulously carved to present various postures and expres-sions they are beloved by the people.

Bass relief patterns on the front, right and left sides of the drum-shaped and square gates' piers are also rich and colorful and uniquely carved. Gates piers vary in size and standard according to the ranks of house owners. 1. Gates piers of Princes' mansions are 0.80metres tall, 0.45 metres wide and o. 36 metres thick. 2. Piers of Guang Liang gates are 0.75 metres high, 0.45 metres wide and 0.30 metres thick .3.Piers of Ru Yi Gates are 0.60 metres tall, 0.29 metres wide and 0.19 metres thick. 4.Gates piers of common people's dwellings are 0.59 metres tall, 0.25 metres wide and 0.19 metres thick. Gates piers of princes' mansions and common people's houses have great difference not only in carving patterns but also in carving workmanship. The bulk of bass relief carvings are flowers such as plum blossoms, chrysanthemums, orchids and green bamboos which are all China's well-known traditional flowers. Also there are carved pomegranate, bat, jade flute, grape and RuYi (a S-shaped article symbolising good luck). A design of musicians and dancers of ancient China carved in bass relief was found on the front and two sides of gate piers of a Prince's mansion in the western part of Beijing. Graceful features and vivid dance movements provide people with aethetic beauty. In some age-old alleys one can see the pattern of a bat holding an ancient coin in its mouth im-plying that good fortune is right before you for in Chinese pronounciation of "钱" Qian or coin and "前" Qian or before is homonymous.

The pattern also symbolises rolling in wealth, the good wish of house owner. If you look carefully one can still find these patterns in today's alleys. But we found out in the course of investigation that very few pairs of stone lions on gates piers were well preserved. Most of the stone lions were damaged during the cultural revolution, Some gate piers have only one intact stone lion left or their pair of stone lions are badly scarred with incomplete heads or tails. The tiny number of well-preserved carved little stone lions in Beijing alleys are treasured works of art indeed.

一对雕刻古老祥云图案的抱鼓石，至今保存完好。
Drum Stones with carved clouds.

雕刻石狮和麒麟图案的汉白玉抱鼓石

Drum Stones with carved stone lion and pattern of unicorn

“马上封侯”图
The design of monkey and horse meaning the future of promotion or granted title

“五福捧寿”图
The design,“Wu Fu Peng Shou”(five bats holding the Chinese character “ 寿” meaning longevity).

“蝙蝠叼磬”图
The bat and Qing (a kind of music instrument used in ancient China)

“三羊开泰”图
The design of three goats meaning auspicious beginning of a new year

雕刻“招财进宝”图案的抱鼓石
Drum Stones with carved the pattern
of Zhao Cai Jin Bao(it symbolizes wealth)

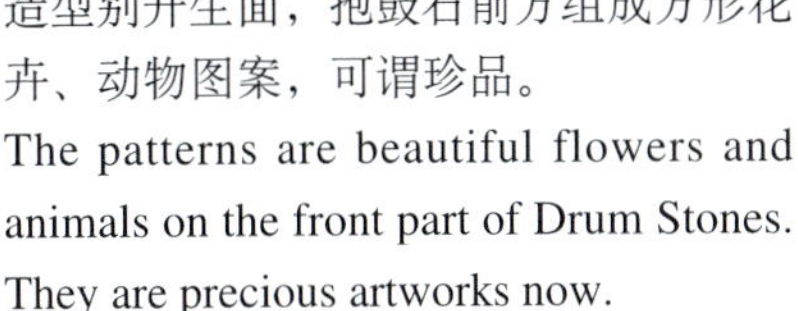

造型别开生面，抱鼓石前方组成方形花卉、动物图案，可谓珍品。

The patterns are beautiful flowers and animals on the front part of Drum Stones. They are precious artworks now.

抱鼓石图案各异，小石狮生动可爱。

There are different patterns of Drum Stones being like vivid little lion.

抱鼓石两侧面雕刻花卉、动物图案。
The patterns of flowers and animals on the both sides of Drum Stones.

古老抱鼓石上的“麒麟图”
The pattern of “Unicorn” on the old Drum stones

抱鼓石上雕刻精美的“奔马图”
Drum stones with carved the pattern of “Galloping Horses”

一对有着 300 年历史的、造型独特老门墩。
The old gate piers has 300 years long history.

不同年代的方门墩

The square gate piers in different period

京城仅存的一对六边形门墩
The only preserved a pair of gate piers of hexagon in Beijing

古老的吉祥人物门墩
左边童子手持莲花图案
右边道童捧太平花图案
The auspicious pattern of figure on gate piers
The pattern of a child with a lotus flower in his hand on the left gate piers
The pattern of a child holding a flower in his hand on the right gate piers

雕有单钱和双钱图案的方形门墩儿（又称福在眼前）。

A square gate pier skulpted with one coin or two coins entitled, “Fu Zai Yan Qian” or happiness is at hand.

雕有龙云图案的方门墩
The pattern of clouds and a dragon on a square gate pier

“五福捧寿”图案的古老门墩儿
The design,“Wu Fu Peng Shou” or five bats holding the Chinese character“寿”(meaning longevity)carved on an age-old gate pier.

不同珍贵图案的方形门墩
The square gate piers of different rare patterns

“狮子绣球”浮雕图案，画面生动活泼，是胡同中仅存之精品。
The vivid and lively carved design, “Lions playing with a ball” carved in bass relief is the only exquisite piece of art so far extant in Beijing alley.

雕有小石狮的方形门墩儿
A square gate pier carved with little lions

方门墩儿，两侧面均雕刻“暗八仙”图案。

The square gate piers are sculptured with designs of 8 immortals presented in a concealed way.

方门墩，两侧面均雕刻“暗八仙”图案。

The square gate piers are skulpted with designs of 8 immortals presented in a concealed way.

雕刻“佛八宝”图案
The patterns of the eight Buddhist treasures symbolized auspiciousness

“奔马图”石雕图案生动，是方形门墩儿中的珍品。它已存世 300 余年。
The design, “Galloping horses”, is a precious work of art found on square gate piers. It has a history about 300 years.

年代久远、图案丰富多彩的方门墩。

Ancient and richful patterns of square gate piers.

不同年代、雕刻不同图案的方门墩。
The square gate piers of different patterns in different periods.

门上的装饰物
——门簪、门钹、门联

Clasps, Cymbal-Shaped Knockers and Couplets —— Decorative Objects on the Gates

门簪是四合院门楼大门上的木制装饰构件，同时起到加固门框的作用，广亮大门上四个门簪，如意门上多为两个门簪，迎面浮雕花卉牡丹、葫芦以示富贵之意。但雕文字的为多数，四个门簪上迎面各雕一字，如“惠我迪吉”其意是引我来至吉祥的地方。两个门簪雕有“吉祥”、“平安”、“迪吉”、“福寿”等字样显示宅主人的向往和愿望。

门钹，由铁或铜材料所制，装饰在大门的左右各一个，成对称位置，其形状类似民乐中的“钹”，称为“门钹”。在其中心装有树叶形金属片或金属圆环。自明、清两代，以至民国，用来敲门发出响亮的金属声，传至院里，当人听到会来开门的。按宅第之分，门钹造型、尺寸大小均有所区别。

门联。北京四合院门楼前不仅有雕刻精美的门头砖雕和门墩，还有大门上的门联。(左边为上联，右边为下联)。门联内容表达了宅主人的思想情操和对美好生活的志趣。精心雕刻在大门上的对联分别采用了隶、篆、行、楷多种书法形式。如果细心阅读胡同里的门联，则既品味了中国传统文化的内涵，又得到了书法艺术上的享受。多数门联为四字、五字、七字对句，但也有少数门联采用三字、八字对句。其内容十分丰富，有的还用黑漆描饰。三字对联如：上联“仁由义”，下联“德载福”。四字对联如：“总集福阴，备致嘉祥”。五字对联有：上联“忠厚传家久”，下联“诗书继世长”。七字对联有：上联“传家有道为存厚”，下联“处世无奇坦率真”。

值得一提的是下门坎。它是门框四边的组成部分，起着坚实门框、容纳门扇的功能；还起到防风、防水、防盗和阻止财气外流的作用。

Clasps are decorative wooden parts on the gates of quadrangles. They also help to reinforce the gates’ frames. Usually Guang Liang gates have 4 clasps and Ru Yi gates 2 clasps. They are carved with peonies or gourds symbolising wealth and nobility. However most of them are inscribed with words. On the top of each of the four clasps is a word. Four words form a phrase like “Hui Wo Di Ji” meaning leading me to the auspicious place. Words carved on two clasps imply “Ji Xiang”(auspiciousness), “Ping An”(plain sailing),“Di Ji”(propitousness) and Fu Shou (good fortune and longevity) expressing the house owners’ aspirations and wishes.

Cymbal-Shaped Knockers are made of iron or copper which are symmetrically fitted on the right and left sides of the door leafs. They are so called because they look like cymbals of Chinese traditional instruments. Ever since the Ming and Qing dynasties down to the Republic of China from 1912 to 1949 they had been used to give out loud metallic sounds. On hearing the sounds people in the quandrangle would come out to open the gate. The knockers differ in design and size in keeping with the type of gate.

Couplet on door panel

In front of the gate tower of Beijing quadrangle, there are not only exquisite the brick carving on the gate head and gate pier, but also couplet on door panel. (the first line of a couplet is carved on the left door leaf, while the second line of a couplet on the right door leaf.) The content of the couplet represents the idea and sentiment of master of the residence as well as his aspiration for inviting life. Manifold calligraphy form. such as official script, seal script, cursive script and regular script was applied on the couplet, which was carved on the gate with meticulous care. If you read the couplet on door panel in the alley with circumspection, you will obtain enjoyment on the calligraphy while tasting the meaning of the Chinese traditional culture. Most of the couplet on door panel is in the form of four -word, five-word or seven-word distich, while some few in the form of three-word or eight-word distich. The content of the couplet is very abound, some of which is painted and decorated with black paint. The three-word couplet, taking for example, the first line of which is "Ren You Yi"and the second line is"De Ji Fu".(means that Benevolence spring from righteousness, virtue brings happiness.) The four-word couplet for instance,"Zong Ji Fu Yin" written in the the first line while "Bei Zhi Jia Xiang" on the second line, meaning that inviting goodness and auspiciousness, collecting happiness and blessing. Taking a five-word couplet for example, it reads "Zhong Hou Chuan Jia Jiu","Shi Shu Ji Shi Chang", meaning that loyalty and honesty carry forward family traditions long, education continues family traditions for generations to come. One seven-word couplet, the the first line of which reads "Chuan Jia You Dao Wei Cun Hou" and the second line of which reads "Chu Shi Wu Qi Tan Shuan Zhen", means that, the way of carrying forward family traditions is none other than to be faithful and honesty, and the philosophy of life has no secrets but to be truthful and sincere.

What's worth mentioning is the lower threshold. It's a component part of the door frame. It serves to reinforce the door frame and tighten the closed door leafs as well as to prevent wind and water, robbery and run off of wealth.

门簪木雕“迪吉”字样，
其意是我来到了吉祥的地方。
Chinese characters “Di Ji” are carved on a wooden gate clasp. Di Ji means auspicious places in English.

配有一对木雕门簪
A wooden gate fitted with a pair of clasps

福星高照

不同造型的铜质门钹
Bronze Cymbal-Shaped Knockers of various shapes

不同造型的铁质门钹

Iron Cymbal-Shaped Knockers of various shapes

门联

The couplet

人傑共和時
物華民主日

迎春接福
兄弟睦家之肥
子孫賢族將大

方形的顶门石墩儿
Square stone piers are shown in the inset

圆形的顶门石墩儿，放置在大门里，关门后放好，以保安全。
Round stone piers are placed inside the gate and are used to reinforce the closed gate to ensure safety.

不同造型的顶门墩
Different shapes of stone piers

步入门楼内见到的木制装饰，花样丰富。俗称其为：倒挂楣子。
“Dao Gua Mei Zi” (different styles of wooden decorations under the back eaves of gate buildings).

珍贵的“上、下马石”与“拴马桩”

Stepping Stones for Mounting or Dismounting from a Horse and Posts for Tethering Horses

在今天的国子监街（原称：成贤街）路北有两座石碑。上面雕刻汉、满文“官员人等至此下马”。由此使我们联想到清朝皇家的规定：满洲贵族（王、贝勒以下）年未满60岁的都骑马往返，汉族官员准许乘轿。这样京城内满族官宦人家及富户子弟出门办事都要骑马，至今一些老胡同里大四合院门前还保留着“上、下马石”和“拴马桩”。门前两侧的一对大青石，一般正面宽0.90m×0.70m或0.70m×0.60m不等，高约0.70m或0.60m，其石底部加出相连的与面宽相等的台阶，这即是“上、下马石”，或称“上马石”。门第有别，规格上也有不同。如：王府前的上马石规格高大，正面雕有花卉，四周部分精刻优美花纹。一般官宦富户门前的上马石，尺寸较小，仅雕光滑平面，不加图案。在门楼往西临街的“倒座”房屋墙上，均装设着四个至六个“拴马桩”，其式样是外周石雕方洞，里边的木柱上装有大铁环，即为拴系马缰绳所用的。今天经过走访考查，典型的大四合院，尤其是门前设有 “上马石”及“拴马桩”的已极难见到了。目前，仅存的几处（指门前设有“上马石”及“拴马桩”的）四合院，也保留的不够完整，有 “上马石”的宅院，已失去了“拴马桩”，有“拴马桩”的宅院门前，却失去了“上马石”。书中选了抢拍下的“上马石”与“拴马桩”实物照片，是很有研究价值的石雕珍品。

On the northern side of today Guozijian jie (street) (originally called Chengxian jie) stand two tablets on which are inscribed the words, fficials and others dismount here in the Han and Manchu languages. This re-minds us of the Qing dynasty regulations governing the imperial family. All Manchu aristocrats (below the rank of prince) under 60 in age should take horse-riding as their means of transport. Han officials were allowed to be carried in sedan chairs. Wherever they went out on errands people of Manchu official or wealthy families in Beijing had to ride on horseback. That why even today some big quadrangles in the alleys still retain stepping stones for mounting and dismounting from a horse and posts for tethering horses. A pair of large blue stone slabs are placed in front of the gate with their front sides measuring 0.90 metres × 0.70 metres or 0.70 metres×0.60 metres in area and a height about 0.70 metres or 0.60 metres. At their bases are added steps with the same width. They are stepping stones for mounting or dismounting from a horse. They differ in standard according to the rank of their master. For instance, the stepping stones for mounting or dismounting from a horse in front of princes residences are big and tall with floral carvings on the top and beautiful, exquisitely-carved designs on four sides, while those of official or wealthy families are smaller in size. Though the stone slabs are of top grade and their surfaces are polished smoothly, they have no carvings of any design. On the wall of the Dao Zuo house facing the street, there are four to six posts for tethering horses. They are wooden posts equipped with big iron hoops which set inside square holes surrounded by carved stones. Through on-the-spot investigation we could rarely find typical big quadrangles with stepping stones for getting on or off from a horse and posts for tethering horses. What the few existing ones have been preserved are incomplete, some have retained stepping stones for getting on or off a horse but lost the posts for tethering horses or vice versa. The real objects whose pictures we took in time and whose pictures were carried in this book are precious stone carvings of great research value.

北京东城国子监胡同里，刻有“官员人等至此下马”汉、满文字之石碑。

In the Guozijian Street in the eastern district of Beijing stands a stone tablet inscribed with the words, “Officials and others dismount here” in Han and Manchu languages.

位于故宫东门外的下马碑

The stone tablet at the easten entrance of the Forbidden City

至今保存着有上下马石及拴马桩的一座四合院（在东城区飞龙桥胡同内）。

A quadrangle at the Feilongqiao Hutong (Alley) eastern Beijing has preserved stepping stones for mounting or dismounting from a horse and posts for tethering horses.

保存完整一对上下马石的四合院门楼

A pair of stepping stones for mounting or dismounting from a horse which have been kept intact

清代时，宅主人出门办事，由仆人牵马，主人登此石上马。如插图中的情景。（李海川　绘）

During the Qing dynasty when a house owner went out on an errand he mounted the horse from these stepping stones while a servant held the reins as shown in the inset. (Illustrator:Li Haichuan)

北京尚存年代悠久的上马石近照。正面浮雕麒麟，侧面浮雕奔马图案。

Time-honored stepping stones for mounting or dismounting from a horse now extant in Beijing.

Bass relief carved unicorn on the front side.

Bass-relief carved galloping horses on the side face.

这是石雕古钱图案造型的拴马桩的近照
A close-up of a post for tethering horses with a stone carving design of ancient coins. This is another form of posts for tethering horses.

影　壁
Screen Walls

影壁，又称为照壁。其位置在四合院门楼的对面，用于遮挡对面杂乱的建筑物，使人们由院内走出大门时感到宽阔、整洁，以示内外有别。古代风水学中，认为影壁是针对冲煞而设置的。《水龙经》云："直来直去损人丁"，古建筑设计中忌讳直来直去，故门楼前设置影壁，或院内设影壁。使气流绕着影壁而行，气则不散，符合"曲则有情"的原理。北京胡同中四合院门前的影壁由砖、石砌成，其造型规格尺寸各有不同，与宅主人的官阶、身份有关。大至可分如下几类：

1．呈"一"字形的，俗称"一字影壁"。其建筑分三段式：上有筒瓦、中壁做出仿木结构的梁框架、框心及四角加上砖雕，下有须弥座。

2．整体造型呈"⌒"形的影壁，名为"雁翅影壁"。砌工精细，磨砖对缝，在影壁墙上边角雕有简单的花草浮雕，使人们看着既庄重且美观。今天在胡同中保存完好的已经不多了。

3．在一些古老胡同里的府第之家，宅门东西两侧，砌影壁墙与门楼檐口成130°左右的夹角，称为：八字影壁（俗称：八字粉墙）。在两侧墙面雕有对称图案的巨幅砖雕。宅门与影壁墙之间留有空地，并放置上下马石一对。

4．北京有些老街巷里四合院门前影壁独特而出名。如：东城的麒麟碑胡同，原有座2.7m余宽的汉白玉影壁浮雕，画面上麒麟人面鳞甲，生有十根角，形态逼真。此物原系明嘉靖年间武将仇鸾府前影壁所嵌，后曾埋于地下，清末时出土，放置在此。胡同也因此影壁浮雕麒麟而传名。辛亥革命后原石雕移至北京的鼓楼内。德胜门内的铁影壁胡同，原留有元代影壁一座，可算北京最古老的影壁，由一巨大火成岩雕制而成，色泽铁黑，故人们称它为铁影壁。现该影壁被移至北海公园快雪堂前。佟府夹道胡同，今166中学，原是清康熙皇帝内亲佟国纲、佟国维的府第。门前原有影壁一座，上镶嵌汉白玉巨石一块，高2.2m、宽0.90m。这块玉石上，有青、褐色天然花纹，雨后经水冲洗，隐约可见在山峦云雾之上，有一观音菩萨坐像，其发髻、披巾、眉眼均由天然花纹形成。坐像前边有一个香炉，香烟缭绕，很是逼真，可称为世间的奇石。这里的影壁因有这观音像画石而传名。此石现保存在该校院内。阜内大街159中学门前有一座巨大影壁，高约5.5m、长约34m、厚1.3m，可为北京影壁中最大的。该址原为历代帝王庙，为宗庙建筑，不可与四合院并列，是北京现存街巷中最长的影壁了。参阅书中提供的照片，定会引起您的欣赏、研究兴趣。

Screen walls called Ying Bi or Zhao Bi or in Chinese are built facing the gates of quadrangles and used to shut out the disorderly buildings outside and to give people a feeling of spaciousness and tideness when coming out of the gates. According to the theory of geomancy screen wall is constructed to dispel evil spirits. The book Shui Long Jing of ancient China says, traightforwardness will reduce the family number". So straightforwardness is a taboo in ancient Chinese architectural designing. That's why screen wall is built in front of the gate or in the courtyard so as to make air flow circle the screen wall in conformity with the principle, "winding arouses interests". Screen walls facing the gates of quadrangles in Beijing alleys are generally built of brick and stone. They vary in design, standard and size according to the offical rank and social status of their owner. They can be roughly divided into the following types:

1. Those shaped like "一" are called Yi Zi screen walls. They have 3 parts: the upper part is covered with semi-circle tiles, the middle part is an imitated wooden structure with brick carvings on four corners and the lower part is the base.

2. Those shaped like "⌒" are called Yan Chi or wild goose-wing shape screen walls. They are meticulously constructed with finelyjoined polished bricks and simple bass relief carvings of flowers and grass on corners, looking sedate and elegant. Not many of them can be found in today's alleys.

3. Ba Zi screen walls, so named because they are shaped like the Chinese character "八". They are erected on the east and west sides of the gates forming an included angle of 130° or so against the gates's eaves. Sometimes each of the two walls of these screen walls bears a similar giant brick carving. Between the screen walls and gate there are open spaces where stepping stones for mounting and dismounting from a horse are installed.

4. Some old Beijing alleys are famous for their unique screen walls. For instance, the Qilinbei or Unicorn Tablet Alley in the east part of Beijing. It formerly had a white marble screen wall over 2. 7metre long carved in bass relief an unicorn with a human face, a body covered with scales and 10 horns. The carving was originally inlaid on a screen wall in front of the residence of Qiu Luan, a general of the Jiajing reign of the Ming dynasty. Later it was buried underground. After being unearthed at the end of the Qing dynasty it was placed in the alley, hence the Qilinbei or Unicorn Tablet Alley. The original stone carving was moved to Gu Lou or the Drum's Tower after the Revolution of 1911. The Tie or Ironcolored Screen Wall Alley inside the Desheng Gate is so named because it formerly had an iron-colored screen wall bearing a giant carving of igneous rock. Buit in the Yuan dynasty it is Beijing' s oldest screen wall. It has now been moved to the front of the Kuaixue Hall of Beihai Park. The No. 166 High School in Jiadao Alley was formerly the mansion of Tong Guogang and Tong Yuanwei, relatives on the queen's side of Emperor Kangxi of the Qing dynasty. Facing the gate was originally a screen wall set with a big piece of white marble 2. 20 metres tall and 0. 90 metres wide. After washed by rains the bluish and brown natural veins of the marble faintly present a statue of Buddha Guanyin sitting above the mountains and clouds. Buddha Guanyin' hair style, shawl, brows and eyes are all shaped by natural veins. It' s really a rare piece of marble in the world. In front of the sitting Buddha Guanyin there used to be an incense burner with curling smokes, making the Buddha Guanyin's image more vivid. Because of this, the screen wall' s name is spread far and wide. This marble screen wall is now kept. In front of the No. 159 High School in Funei Dajie street stands a huge screen wall 5. 5 metres tall, 34 metre long and 1. 3 metres thick. It is Beijing' s largest screen wall so far extant. The site was originally the place where ancestral temples of various dynasties were built. Quadrangles cannot stand side by side with it. The photos carried in this book to which you may refer would arouse your interests to study and appreciate the screen walls in Beijing's alleys.

广亮大门前的照壁（俗称：一字影壁）。
The screen wall in front of a Guangliang Gate. It is popularly known as Yi Zi screen wall or screen wall in a row.

老胡同里，官宦之家四合院前的雁翅影壁。正面成“⌒”形，其砌工精细，磨砖对缝，影壁墙上边角砖雕花卉图案，造型优美。

The wild-goose-wing screen wall in front of the gates of official families. Shaped like “⌒” they are meticulously constructed with finely-joined polished bricks and simple bass relief carvings of flowers on corners.

影壁及局部精美砖雕

The screen wall and the brick carving on it

走进大门里见到的精美照壁

Beautiful screen walls inside the gate of quadrangles

鴻禧

表现民间传说“刘海戏金蟾”故事的图案。
The design portrays a scene from the folk legend “Liu Hai plays with a toad”.

北京市内最长的影壁（位于阜成门内大街路南）
The longest screen wall in Beijing (the south of Funei Dajie street)

位于平安大街上另一个长影壁
Another long screen wall at the Pingan Dajie street

北京 25 片历史文化保护区

The 25 Places under Historical and Cultural Protection in Beijing

20世纪90年代初，北京市做出了两项决定。一是加快危旧房的改造，改善京城百姓的居住条件；二是公布了第一批历史文化保护区。确定南长街、北长街、西华门大街、南池子、北池子、东华门大街、文津街、景山前街、景山东街、景山西街、陟山门街、景山后街、地安门大街、五四大街、什刹海地区、南锣鼓巷、国子监地区、阜成门内大街、西四北一条至八条、东四北三条至八条、东交民巷、大栅栏地区、东琉璃厂、鲜鱼口地区，共计25片。在这些保护区内，有众多的景点和文物古迹，登上景山的万春亭，古老的故宫尽在眼底。当你漫步在那正南正北，东西交错，恰似棋盘的胡同地区，结构独特的四合院门楼呈现在眼前，这是北京人代代生活在这里的天人合一的布局，大、中、小型的四合院建筑是世界大都市中的奇观，这就是北京的胡同。在这些保护地区，都有其悠久的历史和文物景观。东城区的南锣鼓巷是条南北长巷，东西有对称的十六条胡同，建成于元代，历经明、清至今数百年的风云变化，今天依然保存原样。有些胡同的名称，如雨儿胡同、板厂胡同、沙井胡同……至今还延用着元代的名字。其南北主街与对称的胡同组成了蜈蚣形，老百姓称之为“蜈蚣巷”。胡同里不同类型的四合院建筑甚是完整，广亮大门、如意门处处可见。这里保存着清代总督府、皇亲宅第、行辕和名人故居，市级文物保护单位“可园”也在这里。一些古老宅院门楼前还保存着清代的“上下马石”以及雕刻精美的门头砖雕，刻有花卉图案或文字的门簪。大门左右还配有一对石雕的“抱鼓石”或“门枕石”。人们俗称它为“门墩儿”。其造型独特，上面雕有小石狮，多数被砸掉了头尾，这是百年以上的活化石，都有着它自己的故事。门楼前的古槐树是四合院历史的见证。浏览南锣鼓巷地区，不仅可以欣赏、考证四合院建筑，还可以了解胡同文化及民俗。西城区的西四北一条至八条地区，至今保持着明、清二代的格局，四合院保存完整。北头条原名驴肉胡同。北二条清代为帅府胡同、北三条名报子胡同……，至1965年才改称北几条。这些老胡同里，保存着清代四合院多处，还保留着完整的门楼、照壁和垂花门。有的院落里两侧是抄手游廊，墙上建有什锦灯窗及精美的砖雕图案。北三条胡同还有一座带花园的四合院，院内爬山游廊及太湖石相映生辉，真可谓清代四合院的典范。该地区还有名人故居，著名京剧表演艺术家、“四大名旦”之一的程砚秋先生故居即在这条胡同里。这些地区还有多处文物保护单位，以上仅举两片地区为例。今天，25片保护地区历史风貌犹存，各具特色。

读者如有空暇，浏览这些历史文化保护区，置身于胡同古韵中，既开阔了眼界，又可以对老胡同进行研究、考证，增添对胡同文化及民俗的了解。

The first block of historical and cultural site under protection in BeijingTwo decision on old city area was made by Beijing government in the beginning of 90s-: One is to accelerate the reconstruction on dangerous and worn building to improve the living condition of Beijing citizens. The another is proclamation of the first block of historical and cultural site under protection including 25 places: the South Long street, the North Long Street, Xihuamen Dajie(street), Nanchizi, Beichizi, Dong Hua Men Dajie, Wenjin Jie, Jingshan Qian Jie, Jingshan Hou Jie, Jingshan Xi Jie, Dou ShanMen Dajie, Dianmen Dajie, Wusi Dajie, Shishahai, Nanluogu Xiang, Guo Zi Jian, Fuchengmennei Dajie, XisibeiYitiao to Xisibei Batiao(Xisi I alley to the VIII alley), Dongsibei Santiao to Dongsibei Batiao(Dongsi III alley to Dongsi VIII alley), Dongjiaomin Xiang, Dashilan, Dong Liu Li Chang, Xian Huo YU(fresh fishes market). Among those protection area, there are many scenic spots culture relics.

Mounting on the Wanchun Pavilion, one can enjoy the ancient Imperial palace in the bird's eye. Wondering among the chessboard-constructed areas consisting of alleys from east to west or from the east to west, you can spot uniquely-designed quadrangle gates. Those large, medium and small-scaled quadrangle construction, which accommodate the Beijing people for generations, constructed the layout of great harmony between people and heaven- alley. Those preserve areas boasted time-honored historical and culture scenic. Nanluogu Xiang, which located in the Dongcheng district, is a long alley from north to south. Taking Nanluogu xiang as axis, 16 symmetrical alley was arranged along it. Despite hundreds years of weathering and changeable situations throughout the Ming and Qing dynasties, those alleys still remains their original appearances. Some of them, such as Yuer Hutong, Banchang Hutong, Shajing Hutong... still keep the name originated from Yuan dyansty. The south- north oriented main street and the 16 symmetrical east-west oriented alleys present the shape of " Wu Gong (centipede, a kind of reptile), hence popularly known as the Centipede Alleys. The alley boast rows of neat quadrangles with gates of varied styles, such as Guanglinag Gates and Ru Yi Gates. The mansion of viceroy in Qing dynasty, the residence of the imperial relationships, thrill and former residence of celebrities are preserved here. Ke Yuan, the protect unit of Beijing city, is located here as well. Some age-old dwellings still preserve Qingdynasty stepping stones for mounting or dismounting from a horse in front of their gates, gate heads carved with exquisite patterns and gate claps decorated with wood carvings of flowers and words. On both sides of the gates there are well-sculptured round drum stones or rectangular gate pillows. The stone lions carved on the uniquely-designed stones, most of which are damaged with head or tails cut, all have stories behind them. The old locust trees in front of the gates are the long history withness of quadrangle. While visiting the alley area around Nanluogu Xiang, tourists can not only appreciate and research the quadrangle construction there but also get some understanding of the folk customs and culture in alleys. In the western part of the city, the time-honored alley area from the First Alley to the Eighth Alley still retains the original layout of the Ming and Qing Dynasties with quadrangles well preserved. As far as the alley are concered, it was not unitl 1965 that their original names in the Ming and Qing Dynasties were changed to the North Xth (1st, 2nd, 3rd ,...) Alley. For example, the North First Alley , the North Second Allet and the North Third Alley were originally named Lurou (Donkey Meat) Alley, Shuaifu Alley and Baozi Alley respectively (Shuaifu was the official residence of a commander in chief, and baozi refers to an official announcement of a candidates' success in the imperial examinations or of an official's promotion in ancient China.). The alleys here boast many well-preserved quadrangle of Qing dynasty, gates, screen walls and doors with flower-dangled. After entering some of the quadrangle, you can find a Chuihuamen or Dropping Flowers Gate, on each side of the gate there is a short-cut corridor with a row of open windows of different styles bearing a variety of elegant brick carving on the wall. Many dwellings have been rated as historical and cultural site under protection, here listed are two examples: NO.39 in the alley is the former residence of Cheng Yanqiu, a famous artist of Beijing opera performance and one of the four masters playing female roles in Beijing opera. In the Third Alley, a classical 3-layer-courtyard quadrangle with a garden, bearing a splendid combination of a winding veranda and taihu rocks (boulders found on the edge of Lake Tai), may be rated as a model of quadrangles of the Qing dynasty. Today, 25 historical and cultural site under protection retains their historical appearance with respective characteristics.

If readers have vacation, should visit those historical and cultural site under protection in order to widen your knowledge, do research on ancient alleys, get more understanding on the culture, folk customs of alleys and enjoying yourself in the rhythms of ancient alleys.

后记 Postscript

北京胡同的四合院门楼建筑是非常独特的。门楼造型、砖雕、木雕、石雕装饰诸方面都有着研究和观赏价值。经历元、明、清至今，北京仍保留着大量的胡同以及一座座优美的、建筑风格各异的四合院门楼。

北京这座古城，在现代化高速发展的时刻，老胡同地区在进行危改和拆迁。这里发表的大量照片及插图对研究胡同建筑文化有着一定的参考价值。其中的部分照片可谓精品。

近年，北京市政府公布了第一批历史文化保护区。书中有一篇文章专门介绍了这些地区的建筑与历史，以便读者参观考察时参阅。

The construction of quadrangle gate building in the Beijing Hutong (alley) boasts their special originality. There are rich value of study and appreciation embedded in their shape, brick carving and stone carving decoration. After undergoing the dynasty of Yuan, Ming, Qing, a large number of Hutongs has been reserved to the present. The delicate gate building, having experienced many years of swift changes of the world, still preserve a multitude of fine art article. Beijing, the historic city experiencing modernization high–speed development, is witnessing its aged alleys and region under reconstruction of old houses and rehousing. The volume of photos and a few illustrations released in the book has reference value for study of the construction and cultural of Hutong, some of which can be called collector's item.

In recent years, the Beijing city government has promulgated the first batch of historial and cultural conservation zones. One article in the book is specialized to introduce the building and history of the zones for the reference of tourists.

图书在版编目(CIP)数据

胡同门楼建筑艺术／李明德著．—北京：中国建筑工业出版社，2003
ISBN 7-112-05518-0

I.胡... II.李... III.民居—建筑艺术—北京市—图集 IV.TU241.5-64

中国版本图书馆CIP数据核字（2002）第087767号

翻译：严爱榛　武嘉澍　李海川
插图：李海川
责任编辑：时咏梅
版式设计：刘向阳

胡同门楼建筑艺术
李明德　著
李海川　摄影

中国建筑工业出版社出版、发行（北京西郊百万庄）
新　华　书　店　经　销
北京画中画印刷有限公司印刷

开本：889×1194毫米 1/20 印张：6 字数：170千字
2003年4月第一版 2005年1月第二次印刷
印数：2,001-3,200册 定价：50.00元
ISBN 7-112-05518-0
TU·4848(11136)

（邮政编码 100037）
本社网址：http://www.china-abp.com.cn
网上书店：http://www.china-building.com.cn